SNAKES

CORN SNAKES

by Julie Murray

Cody Koala
An Imprint of Pop!
popbooksonline.com

Hello! My name is Cody Koala

This book is filled with videos, puzzles, games, and more! Scan the QR codes* while you read, or visit the website below to make this book pop.

popbooksonline.com/corn

*Scanning QR codes requires a web-enabled smart device with a QR code reader app and a camera.

abdobooks.com
Published by Pop!, a division of ABDO, PO Box 398166, Minneapolis, Minnesota 55439.

Printed in the United States of America, North Mankato, Minnesota.

042025
082025

THIS BOOK CONTAINS RECYCLED MATERIALS

Cover Photo: Shutterstock Images
Interior Photos: Adobe Stock Photo; Getty Images; Shutterstock Images; Science Source
Editor: Elizabeth Andrews and Tyler Gieseke
Series Designer: Neil Klinepier, Colleen McLaren

Library of Congress Control Number: 2024948404

Publisher's Cataloging-in-Publication Data
Names: Murray, Julie, author.
Title: Corn snakes / by Julie Murray
Description: Minneapolis, Minnesota : Pop!, 2026 | Series: Snakes | Includes online resources and index
Identifiers: ISBN 9781098247805 (lib. bdg.) | ISBN 9781098248345 (ebook)
Subjects: LCSH: Corn snake--Juvenile literature. | Red rat snake--Juvenile literature. | Constrictors (Snakes)--Juvenile literature. | Snakes--Behavior--Juvenile literature. | Snakes--Juvenile literature.
Classification: DDC 597.96--dc23

Table of Contents

Chapter 1

Where Do They Live?

Corn snakes are beautiful **reptiles**. They live in the southeastern United States and northern Mexico. They are not **poisonous**.

Watch a video here!

Corn snakes are active at night. During the day, they are found hiding under logs, in animal **dens**, and in old buildings. They are good climbers and can be found in trees, too.

Where Corn Snakes Live

Corn snakes live for about seven years in the wild.

Many people like to keep corn snakes as pets. Corn snakes are peaceful and easy to care for. They live in special **habitats**. They can live up to 25 years in human care.

Chapter 2

What Do They Look Like?

Corn snakes are skinny and grow 3 to 6 feet (0.9–1.8m) long. They can weigh 1 to 3 pounds (0.45–1.4kg). A corn snake's body is covered in shiny **scales**.

Learn more here!

A corn snake can be red, brown, orange, or yellow. It has a marking on its head that looks like a spear.

It also has reddish-brown **saddle**-shaped markings. The markings are outlined in black.

A corn snake's tummy has black-and-white checkerboard markings. This is how the snake got its name. Its tummy looks like pieces of corn.

Adult corn snakes shed about four times a year.

Like other snakes, corn snakes shed their skin. Their skin and eyes turn cloudy before shedding. They slide out of their skin, leaving it in one piece.

Chapter 3

How Do They Hunt?

Corn snakes are constrictors. They grab their **prey** with their teeth. Then, they wrap around it and squeeze. They eat their prey whole. They eat mice, birds, frogs, and other snakes.

Explore links here!

Chapter 4

Baby Corn Snakes

Female corn snakes lay 10 to 30 eggs in a **den**, tree stump, or pile of plants. The eggs hatch 60 days later. The babies are 10 to 15 inches (25 to 38cm) long.

Complete an activity here!

Making Connections

Text-to-Self

Corn snakes are popular pets. Would you want to have a corn snake as a pet? Why or why not?

Text-to-Text

Have you read books about different kinds of snakes? If so, how are those snakes similar to or different from corn snakes?

Text-to-World

Corn snakes are constrictors that squeeze their prey. Can you think of other snakes that are constrictors?

Glossary

den – a safe resting place for wild animals, often underground.

habitat – for pets, a comfortable, safe, and secure space to live that is like their wild environment.

poisonous – filled with or containing poison.

prey – an animal that is hunted by other animals for food.

reptile – a cold-blooded animal with a skeleton inside its body and dry scales or hard plates on its skin.

saddle – a leather seat that is used on the back of a horse to carry a rider.

scales – small, hard, thin plates that cover reptiles.

Index

Online Resources

popbooksonline.com

Thanks for reading this Cody Koala book!

This book is filled with videos, puzzles, games, and more! Scan the QR codes* while you read, or visit the website below to make this book pop.

popbooksonline.com/corn

*Scanning QR codes requires a web-enabled smart device with a QR code reader app and a camera.